Great Scientific Questions and the Scientists Who Answered Them

HOW DO WE KNOW THE AGE OF THE EARTH

CHARLES J. CAES

Great Scientific Questions and the Scientists Who Answered Them

HOW DO WE KNOW

THE AGE OF THE EARTH

THE ROSEN PUBLISHING GROUP, INC.
NEW YORK

To my wife Karen, for all her love, patience, and support

Published in 2001 by The Rosen Publishing Group, Inc.
29 East 21st Street, New York, NY 10010

First Edition

Library of Congress Cataloging-in-Publication Data

Caes, Charles J.
How do we know the age of the earth / Charles J. Caes. — 1st ed.
p. cm. — (Great scientific questions and the scientists who answered them)
Includes bibliographical references and index.
ISBN 0-8239-3381-4 (lib. bdg.)
1. Earth—Age—Juvenile literature. 2. Geological time—Juvenile literature. [1. Earth—Age. 2. Geological time.] I. Title. II. Series.
QE508 .C33 2001
551.7—dc21

00-012607

Cover images: layers in rock formations.
Cover inset: trilobite.

Manufactured in the United States of America

Contents

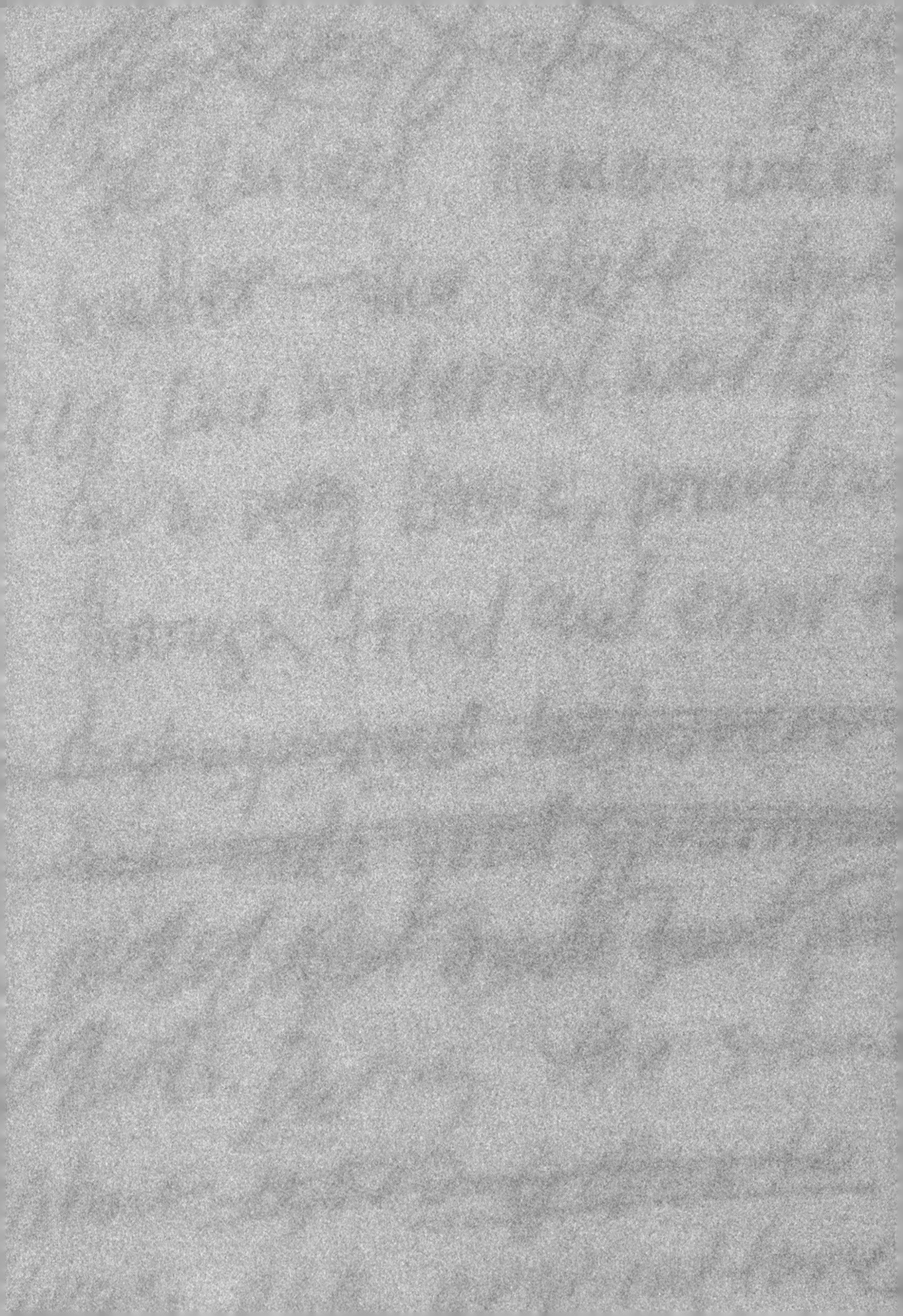

Religion to Physics

How old is the earth? Clearly, it is older than any of the people living on it, and older than all of recorded human history. So there is no one we can ask for an answer. The earth itself will have to provide the answer. The earth is its own artifact, and contains important records of its own past. A

Fossils, such as this ammonite, give scientists an indication of how life evolved and enable them to more accurately estimate the age of the earth.

fossil, for example, is more than just the petrified remains of an organism. It is nature's record of a past life form. Its composition, chemistry, and location tell a story of what its world must have been like. Rocks, too, in terms of their composition and location, tell a story about their own birth and age. Everything in nature serves as a record keeper in one way or another. To discover the age of the earth, then, we have to study the physical and chemical changes that have taken place in the past and the way that these changes have been recorded in the natural world. This is the scientific point of view, and today it seems to us to be the obvious way to settle the question. But this has not always been obvious. Until the eighteenth century, however, estimates for the age of the earth were primarily based on religious ideas or the analysis of sacred texts.

Many ancient cosmologies considered time to be infinite and described the earth as part of an eternal cosmic cycle, during which events occur over and over again. Just as the universe was born by the will of the gods or some cataclysmic event, so will it some day destruct, only to be born again. These ancient religious ideas regard the universe as changing but static, and the difficult questions of beginnings and endings are avoided.

Some cultures attempted to assign dates to these cycles. In ancient Vedic writings, the oldest sacred texts of Hinduism, the world exists for one *kalpa,* a day and a night in the life of Brahma, creator of the universe. Each kalpa is the equivalent of 1,000 *chaturyugies*, each consisting of 4.32 million years. That means that the earth will exist for 4.32 billion years. Using Hindu numerology to count to the beginning of the twenty-first century, exactly 1,960,850,098 years of this time have passed so far.

The Greeks also believed in an eternal cycle, as Plato taught. He believed that this cycle lasted for 26,000 years. By contrast, early rabbinical chronologies place the date of creation at 3,700 BC, and the Babylonians, according to the third century BC priest and historian Berossus, believed that the world was a half million years old. As Christianity took hold, cyclical concepts of creation and destruction lost favor. The coming of Christ and the Last Judgment had to be one-time events, and the idea of an Earth older than biblical chronologies challenged sacred writings.

Among the first to draw upon Christian scripture to estimate the age of the earth was the second century Christian scholar Theophilus of Antioch (115–181). Theophilus calculated an age of 5,698 years for the earth, which, when carried from his time to ours, makes the earth about 7,668 years old. Theophilus admitted that this was just an educated guess, but he felt it was a much more reliable estimate than those of misguided scholars who believed the earth to be tens of thousands of years old. By misguided scholars, he meant the Greek and Roman intellectuals still fascinated with pre-Christian ideas.

After Theophilus, over the next thousand years, more than 200 Christian scholars would come up with a range of estimates for the age of the earth. Among them were Saint Basil the Great (330–379), one of the four fathers of the Greek Church, who estimated 5,994 years; Saint Augustine (354–430), the noted theologian, who arrived at 6,321 years; and Giovanni Pico della Mirandola (1463–1494), a Dominican priest and mystic, who calculated an age of 5,471 years.

These ages were determined by a careful study of the Bible, working backward to estimate the time periods between major events such as the births and deaths of patriarchs and prophets, the reigns of kings, and the time between major events such as wars and the collapse of empires. Some scholars were so sure of biblical authenticity and their own calculations that they felt able to pinpoint the exact time of creation. John Lightfoot (1602–1675), for example, a well-respected biblical scholar, gave the probable date of birth for Earth as 9 AM in the year 3982 BC.

The most influential estimate based on scripture was that of James Ussher (1581–1656), the Irish prelate and scholar. Ussher determined that God created the earth on October 22, 4004 BC. He was so convincing that for centuries King James versions of the Bible footnoted this date as the start of creation. By 1900, however, the advance of science had so challenged the credibility of this date that new editions of the King James Bible published by Cambridge University no longer

Like many scholars of his day, James Ussher used the Bible to estimate the age of the earth.

mentioned it, and other publishers eventually followed suit.

A more scientific view began to emerge very slowly. In 1669, Danish physician Nicolaus Steno (1638–1686) began to make some extraordinary claims about fossils. Official church doctrine held that fossils were either the remains of animals killed in cataclysmic events like Noah's flood, or were the now-abandoned works of God's previous attempts at creation. Steno suggested that fossils were the remains of ancient animals that had lived out their lives normally. There was no need for a religious explanation or a biblical event to explain their presence. Steno also studied the different layers of rock he found exposed in the hills of Tuscany, Italy, where he was employed as a doctor to a Medici prince. He concluded that, no matter how twisted these layers were, they were originally laid down horizontally. The youngest layers of rock were on top, and the oldest layers of rock were underneath. This has come to be known as the principle of superposition. The various layers were composed of different materials and contained different types of fossils, and there were abrupt changes between layers. The whole arrangement suggested a building up of materials over great periods of time. Ironically, having made these revolutionary suggestions, Steno in 1677 became a Roman Catholic bishop and abandoned science for religion.

Among the first to challenge the idea of a relatively young Earth was Frenchman Benoit de Maillet (1656–1738), a government diplomat whose responsibilities gave him ample opportunity to travel about and pursue his interest in geology. De Maillet believed that in the past the earth had been entirely covered by water, proof of which came from the fossil evidence of many marine animals in places that were now dry land. The sea level must have been in continual decline, he reasoned. If he could determine the rate of this decline, he could estimate when the first mountains appeared above water.

De Maillet observed in coastal areas that certain ancient buildings were originally constructed at sea level but were now conspicuously higher. Applying a little math to the age and current elevation of those structures, he determined that the rate of sea level decline must have been about three to four inches every 100 years. Estimating the height of the tallest mountains, he was then able to determine when they were first exposed by the receding waters that he believed once covered the planet. He estimated Earth to be more than two billion years old.

De Maillet's reasoning was hopelessly simplistic. Sea level is determined by rising land masses as well as

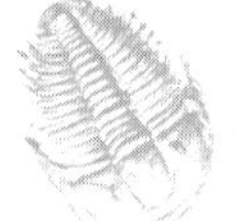

changes to the ocean floor. Earthquakes, erosion, the dynamics of plate tectonics, glaciation, mountain building, and other processes have the effect of halting, reversing, or speeding the decline of sea levels, thereby complicating any efforts to develop a dependable timescale. And the earth was never entirely covered by water. The temperature of early Earth was far too hot to allow the formation of oceans. Only after the planet cooled could clouds of water vapor be released from the sky to form the oceans. Nevertheless, science owes de Maillet a debt. He chose not to look to biblical authority for his answer, and he believed that nature itself might reveal its geological history and that there was sufficient reason to suppose that the earth was much older than biblical chronologies indicated.

Benoit de Maillet used flawed reasoning about sea levels to try to determine the age of the earth. Sea level is determined by rising land masses, such as the San Andreas Fault, and by other geologic and climatic forces.

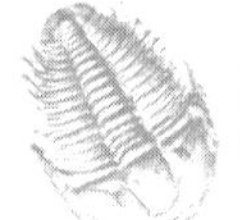

JAMES HUTTON AND UNIFORMITARIANISM

A physical theory of the planet's age began to emerge during the 1770s and 1780s, thanks mainly to the work of Scotsman James Hutton (1726–1797). Hutton had trained as a doctor but never took up a practice. He had earned enough money from a successful business that manufactured fertilizer to retire and dedicate his life to the study of geology. In fact, he is considered the father of geology, since before the publication of his book *Theory of the Earth* in 1785, geology did not exist as a distinct science. The natural history of the earth was considered a subject fit only for religious scholars. Hutton reasoned that scripture might be correct in placing man's origin in the not-too-distant past, but that the natural history of the earth and other forms of life indicated spans of time that were far beyond human comprehension.

Hutton noted the extent to which the earth was covered with fossils. He thought that it must have taken longer than biblical chronologies indicated

James Hutton championed uniformitarianism, which postulates that the earth evolved slowly over time.

for such quantities to accumulate. He also carefully studied the physical features of Britain's landscape and noted the common processes that changed it—erosion from water and wind, the compression of sediments into rock, and materials brought to the surface by volcanic eruptions. He concluded that these same common processes operated in the past, and that given enough time, these processes could explain the appearance of the surface of the earth. The amount of time these forces needed to do their work had to be measured in hundreds of millions of years.

Hutton's ideas came to be known as the doctrine of uniformitarianism, and held that the earth evolved slowly over immense periods of time, subject to forces that were still at work and that wrought change gradually. These ideas were in conflict with the doctrine of catastrophism, the notion that the earth's landscape was altered in sudden, rapid, and dramatic cataclysms. Catastrophism was the theory supported by French anatomist Georges Cuvier (1769–1832) and later by his pupil, Harvard paleontologist Louis Agassiz (1807–1873). Cuvier was a brilliant student of fossils (he discovered and named the flying reptile pterodactyl), but he could not bring himself to challenge biblical teachings. He believed that the fossil record indicated a world periodically inundated by catastrophic floods

French anatomist Georges Cuvier supported catastrophism, the notion that the earth's landscape was altered in sudden, rapid, and dramatic ways, including eruptions such as this one.

that wiped out all previous organisms, after which God would restart creation all over again.

Cuvier's pupil Agassiz maintained his belief in catastrophism even after Charles Darwin had published his theory of evolution, but his reasoning was more scientific. Before coming to teach at Harvard, Agassiz had studied the movements of glaciers in Switzerland and northern Europe, and he was the first to propose that the earth had experienced an ice age. Such a radical change of climatic conditions challenged Hutton's idea of gradual, uniform development, and eventually forced geologists to adopt a modified form of uniformitarianism.

Cuvier supported catastrophism as it did not directly challenge biblical teachings about the age of the earth.

Catastrophism appealed to theologians, who could interpret these sudden dramatic changes as acts of God, like the flood in Genesis. Catastrophism did not require vast periods of time to work, and it did not conflict with biblical interpretations of the earth's age.

Agassiz clung to his belief in catastrophism even after Darwin published his theory of evolution.

Uniformitarianism was a radical challenge to the religious teachings of the day. Hutton's ideas were simplified and popularized after his death by his friend, Scottish mathematician John Playfair (1748–1819), in Playfair's book *Illustrations of the Huttonian Theory of the Earth*, published in 1802, and by Charles Lyell (1797–1875) in his *Principles of Geology*, published in 1833. But Hutton's theory came

After studying glaciers in Europe, Agassiz proposed that the earth had experienced an ice age. This challenged Hutton's idea of gradual, uniform development, eventually forcing geologists to adopt a modified form of uniformitarianism.

to be accepted only gradually, after supporting observations and research by William Smith (1769–1839) and Charles Darwin (1809–1882).

William Smith, a British engineer and surveyor, became involved in canal building during England's industrial revolution. At digging sites he was able to observe the strata or layers of different types of rocks that were uncovered, and he noted that each strata contained its own unique fossils, not found in other layers. Strata, in fact, could be identified by the fossils they contained, no matter how twisted or broken the different layers of earth. Today such fossils are known as index fossils. Like Steno before him, Smith surmised that strata nearer the surface of the earth were younger than deeper strata, so the relative age of the fossils could be worked out according to how deep in the earth they were found. This is known as the principle of fossil succession.

Furthermore, the deepest fossils were the most alien in appearance, and the higher fossils more closely resembled living life forms. "Strata" Smith, as his colleagues called him, had discovered compelling evidence for an evolutionary view of life, twenty years before Darwin, though Smith himself did not realize the implications of his discoveries.

CHARLES DARWIN

In 1831, Charles Darwin, a restless young man who had given up the study of both medicine and theology, and whose father thought that he would not amount to much as a result, joined Captain Robert Fitzroy of the HMS *Beagle* on a scientific expedition that would last five years. Darwin had developed an interest in natural history when he was at Cambridge and had read some of the works of Scottish geologist Charles Lyell, who promoted Hutton's notion of a slowly evolving Earth. Lyell's three volume work, *The Principles of Geology,* published in 1833, had become the bible of the uniformitarians, and Darwin fell under its influence.

Charles Darwin's theories of natural selection and evolution revolutionized how human beings think about and study life and the earth.

Charles Lyell promoted James Hutton's notion of a slowly evolving Earth. His writings influenced Darwin.

As he voyaged along the coast of South America, and particularly as he studied the wildlife of the Galapagos Islands near Ecuador, Darwin noticed how species that were geographically isolated by islands or mountain ranges had distinctive features or behavior patterns. He studied in particular several species of finches that differed with respect to size, the shape of their beaks, and their behavior and feeding patterns, depending upon the habitat they occupied and what they had available for food. It was clear that these species were changing themselves and adapting to their environment. But how could they do this?

Back in London in 1838, Darwin read the famous essay on population by Thomas Malthus (1766–1834), published four decades earlier. Malthus believed that human populations would grow faster than the available food supply, and that competition for food would lead to poverty and starvation. The notion of intense competition among organisms for the environment's resources intrigued Darwin, and he saw in this competition the mechanism for species evolution. Over time, species produced random variations of themselves. Some of these variations gave organisms better ways of finding food or adapting to their environment in other ways. These organisms would survive and reproduce more of themselves, while the less well-adapted organisms would die out. Darwin called this process natural selection. Given enough time, these variations or changes would accumulate and produce creatures very different in appearance from their ancestors. Just as humans could breed new types of animals by artificial selection, nature could induce the same kinds of

No. 85
PLATE CCCCXXIV.
Drawn from Nature by J. J. Audubon, F.R.S. F.L.S.
Engraved, Printed and Coloured by Robt. Havell 1838.
Lazuli Finch.
FRINGILLA AMŒNA, Say
1. Female
Crimson-necked Bull-finch.
PYRRHULA FRONTALIS, Bonap.
2. Male
Grey-crowned Linnet.
LINARIA TEPHROCOTIS, Swains.
3. Male
Cow-pen Bird.
ICTERUS PECORIS, Bonap.
Evening Grosbeak.
FRINGILLA VESPERTINA, Cooper
Brown Longspur.
PLECTROPHANES TOWNSENDI, Aud.
7. Female

changes by placing environmental pressure on organisms to adapt or die.

Darwin published his theory in 1859 in his famous book *The Origin of Species*, and in 1871 he took the bold step of asserting, in *The Descent of Man*, that human beings, too, had evolved from other organisms. Just how species produced variations of themselves was unknown to Darwin. He never learned of the work of Austrian botanist Gregor Mendel (1822–1884), who worked out the laws of heredity, and he died a century before scientists discovered genes and how they mutate. But evolution by natural selection was a powerful idea that held out the promise of explaining the relationship between all living things, and the most advanced scientists of the day eventually accepted it.

While at the Galapagos Islands, Charles Darwin studied several species of finches and found that they were adapting to their environment. He used these observations to develop his theory of natural selection, which explains that organisms must adapt to environmental pressures or die.

The key point of Darwin's theory was that if a human being had evolved from a creature as different as an ape or a fish, vast amounts of time were required for the necessary physical changes to take place. Evolution was a very slow process that matched the gradual geological changes the uniformitarians believed in. It was no accident that Darwin and Lyell became friends and supporters of each other's ideas. To truly understand the time frames involved, Darwin suggested learning about geological processes, the extent to which the land has been sculpted and eroded, and how much sediment has been deposited over time. His work gave strong support to geologists advancing theories of a truly ancient Earth.

The Quest for the Earth's Age

As the implications of the work of Lyell and Darwin were being debated, geologists were searching for new and more scientific methods of measuring the age of the earth. One such method involved studying the formation of sedimentary rocks. Such rocks are formed when layers of sediment—particles of sand, silt, or clay—are deposited and pressed together by the weight of the materials above. By measuring the extent or thickness of sediment deposited during various periods of history and dividing by the rate at which this sediment was deposited, geologists believed that they could measure the time it took for a layer of sedimentary rock to form.

According to the published results of these calculations, the earth's age ranged from 25 million years to billions of years. But the method used to estimate the age of sedimentary rocks was full of inherent problems that rendered the results suspect and, finally, indefensible. To begin with, any method that depended upon accumulations of sediment must necessarily assume a uniform rate of deposition over time. For a geologically active planet like Earth, this was an assumption scientists could not prove and therefore could not make. There were also gaps in the stratigraphic record, missing layers of rock revealed by the great differences in fossil forms found above and below the gap. Did these gaps represent periods in which no sediment was deposited, or periods where materials were eroded away, and how much time was represented by such gaps? There was no way to tell at the time.

A PHYSICIST CHALLENGES DARWIN

An entirely different method of measuring the earth's age was suggested in 1846 by the brilliant young Scottish physicist William Thomson (1824–1907). Thomson was a child prodigy who presented his first paper on mathematics to the Edinburgh Royal Society when he was still a teenager. His primary field of interest was the flow of heat energy, and he eventually made discoveries in thermodynamics that resulted in a knighthood and the title of Baron Kelvin. Thomson assumed that the earth and all the planets were molten hot at the time of their formation. They were originally the same temperature as the Sun and had been cooling off at a steady rate since their

formation. Thomson was aware that in mine shafts the temperature increased the deeper you went, and he assumed that this higher temperature was the result of heat still present from the time of the planet's formation. Thomson performed his calculations and concluded that the earth was somewhere between 20 million years and 400 million years old, and probably closer to 100 million years old.

These numbers created serious problems for Darwin, Lyell, and all uniformitarians and evolutionists. This was, first of all, a scientific rather than a religious argument, and it had to be answered. If Thomson's theory was true, there was simply not enough time for geological processes to shape the earth or for natural selection to generate the biological

Scottish physicist William Thomson (aka Baron Kelvin) challenged Darwin's view of an ancient earth because he believed the planet was still too hot to be very old.

changes that would produce complex organisms. The geologists believed that layers of sedimentary rocks were far too thick to have formed in a mere 20 million years or even in 500 million years. Some geologists and evolutionists were actually considering that the earth might be billions of years old. Thomson, on the other hand, felt that his higher estimates were at least ten times too great, and that no amount of scientific evidence would convince his stubborn opponents that the earth was a young planet.

Other researchers would use Baron Kelvin's methods. Among the more influential was Clarence King (1842–1901), the first director of the United States Geological Survey. In 1893, he published a paper in which he put the age of Earth at only 24 million years. Then, in 1897, Kelvin himself redid his calculations and this time arrived at an age of 20 million to 40 million years. He personally preferred the lower age. One of his colleagues, Scottish physicist and mathematician Peter Guthrie Tait (1831–1901), also did the math and produced even lower estimates than Kelvin, from 10 million to 15 million years.

Baron Kelvin was a recognized genius who had made revolutionary discoveries about the transmission of heat energy, and it was difficult to ignore or disprove his objections to the new theories of geology and evolution that required a very old Earth. But the geological evidence and the fossil records were also compelling, and the Darwinians and uniformitarians held fast to their belief in an ancient Earth, and in the end they were proved to be correct. But it was not until the first decades of the twentieth century that the reasons for Baron Kelvin's miscalculations became evident.

Baron Kelvin, the great expert on heat, did not understand how the Sun burned. During his time, no one knew anything about the processes within the atom or the tremendous amount of energy contained in the atomic nucleus. Baron Kelvin thought that the Sun burned no differently from a very hot fire, and that its temperature was, like the temperature of the molten rocks that formed the early Earth, only a few thousand degrees. Lacking any understanding of the process of hydrogen fusion that is responsible for the Sun's enormous energy output, Baron Kelvin could not conceive of a Sun older than a few hundred million years, because it didn't have the fuel to burn for longer than that time. And if the earth had reached its present state after cooling off from a temperature of only several thousand degrees, it could not possibly be billions of years old. When he studied the temperatures found in deep mines, he had no idea that radioactive elements in the earth's crust were continually giving off more heat, and that the earth had not been cooling at his predicted rates.

For more than fifty years, Kelvin's objections to an ancient Earth were a problem for geologists and Darwinians. But they sensed that they were right and that Kelvin had missed something, and they proceeded to expand their theories and gather evidence in support of those theories from other scientific investigations. In

1899 Irish professor of geology and mineralogy John Joly (1857–1933) attempted to discover the age of the earth by measuring the rate of salt accumulation in the oceans. Two centuries earlier English astronomer Edmund Halley (1656–1742) had suggested that the salt in the sea came from erosion from the land and was deposited in the oceans through river runoff. Joly estimated the salt content of the oceans and the rate of deposition from rivers and came up with an age of between 80 and 90 million years. But his methods were flawed. He had assumed that the sea was a closed system and that salt just kept accumulating. In fact, salt is lost through evaporation and sedimentation, and the salt content of the sea does not increase but is kept in balance by various geological processes.

TIDAL FRICTION

William Thomson/Baron Kelvin, with the usual enthusiasm that kept him at the forefront of scientific inquiry,

Tides are caused by the pull of the Moon. This pull creates a resistant force that works against the rotation of the earth.

now developed a new way of calculating geological age, based on the concept of tidal friction. He began to investigate the way in which the gravitational pull of the Moon influences both the earth and the ocean.

The Moon's gravity pulls the earth toward it as one solid object, as though the planet's entire mass were concentrated at its center. But the water on the

Over time, the rotation of the earth slows, and the lost energy becomes heat that is released into space.

planet is free to move about. This causes the tides, the daily rise and fall of sea level as the oceans come under the influence of the Moon's gravity. The pull of the Moon on the oceans creates a resistance or frictional force working against the rotation of the earth. Over time, the rotation of the earth slows down slightly, and the rotational energy lost is transformed into heat and released into space.

With a slowing of the earth's rotation comes a lengthening of the earth's day. According to Kelvin's calculations, the day should increase by about one-tenth of a second every 5,000 years. Working backward,

Kelvin imagined a time when the earth might have been spinning at twice its current rate, and concluded that under such circumstances the oceans would have been trapped in the polar regions of the planet. As this is clearly not the case, Kelvin believed that the earth must have solidified at a time when its rotational period was not all that different from what it is today, and that therefore the earth was not that old.

Just a few years later, in work that began in 1879, George Darwin (1845–1912), the second son of Charles Darwin, expanded on Kelvin's new method. The younger Darwin was a Cambridge-educated astronomer most well known today for a theory, now out of favor, that the Moon was part of Earth until a cataclysmic event tore it loose. This notion influenced his thinking on how to calculate the age of the earth.

The younger Darwin looked at the effects of tidal friction on the entire Earth-Moon system. He understood that if tides reduced the rotational speed of

Earth, this loss of energy had to be made up somewhere, probably through an increase in the Moon's angular momentum. To increase its momentum, the Moon had to move farther away from Earth. It was therefore much closer to Earth in the distant past. By extrapolating backwards, Darwin tried to estimate how long ago the Moon was in contact with Earth. He made a number of calculations, arriving in 1898 at a minimum age of 56 million years. Today, the effects of tidal friction on Earth's rotation and the Moon's distance from Earth are accepted physical phenomena, but scientists no longer believe that the Moon was once part of Earth. The Moon is believed to have formed independently.

As the twentieth century approached, scientists had established a powerful theoretical framework for an ancient Earth with bold new theories of geological and biological evolution. They had also devised some new empirical methods for measuring the age of the earth. But those methods were still very crude and

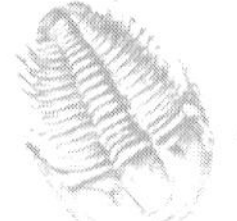

subject to many errors, and none of the methods had shown that the earth was old enough to support the new theories. It must have been a frustrating time for uniformitarians and Darwinians, who sensed that they were right but could not prove it, certainly not to the satisfaction of the physicists. And yet, as the new century dawned, it would be revolutionary discoveries by the physicists that would provide the tools needed to prove the uniformitarians and Darwinians right.

Nature's Clock

In 1896, Frenchman Antoine Henri Becquerel (1852–1908), a professor of physics at the École Polytechnique in Paris, stumbled upon an unusual characteristic of the heavy element uranium. Becquerel was investigating whether or not there were any naturally occurring substances that might produce the kind of X rays discovered by the German physicist Wilhelm K. Roentgen (1845–1923) in 1895. To this end he was experimenting with a uranium compound that he exposed to a photographic plate wrapped in black paper. The result was a fogging of the plate, as if it had been exposed to light. Becquerel had discovered the phenomenon of radioactivity, so named by Marie (1867–1934) and Pierre (1859–1906) Curie two years later when they discovered the same phenomenon in other elements like radium.

The nature of radioactivity was not readily discernible. Becquerel and the Curies did not know exactly what it was, although the Curies soon realized that radioactivity was a property of atoms and was not dependent upon the chemical compound in which it might be found. The mystery would be solved by British physicist Ernest Rutherford (1871–1937) and his young assistant Frederick Soddy (1877–1956).

Rutherford repeated Becquerel's experiment and noted that there were three types of "rays" being emitted by the uranium sample. He labeled these alpha, beta, and gamma rays. The alpha rays eventually proved to be helium nuclei, consisting of two protons and two neutrons. The beta rays carried a negative electric charge and were eventually determined to be electrons. The gamma rays were not particles at all, but highly energetic electromagnetic waves.

Once it was understood that some of the radioactive products were subatomic particles, Rutherford and Soddy put forward a theoretical explanation for radioactivity. Radioactive elements were disintegrating. The atoms of these elements were spontaneously ejecting fragments of themselves and transforming themselves into other elements. The "parent" elements would decay into more stable "daughter" elements. In 1903 Rutherford and Soddy demonstrated that some radium atoms in a radium compound transformed themselves into helium atoms. Rutherford, in fact, was the first person to achieve such a transmutation of elements, dreamed of since the days of the alchemists.

Rutherford and Soddy noted some other interesting aspects of radioactive decay. It was impossible to predict when an individual atom would disintegrate. That atom might emit an alpha particle now, an hour from now, tomorrow, or five years hence, and there was no way to know when this would occur. But the overall rate of decay from a sample of a radioactive element was always constant. A friend of Rutherford's, American chemist Bertram B. Boltwood (1870–1927), demonstrated that uranium decayed into lead, and in 1907 he suggested that it might be possible to determine the age of the earth's crust using uranium's constant rate of disintegration. If one measured the ratio of the amount of the parent radioactive element to the amount of the stable daughter element in various rocks, one could determine how long the uranium in those rocks had been decaying since the rocks were formed.

RADIOMETRIC DATING

The discoveries of Rutherford would have a profound impact on modern physics, but for geologists it meant the discovery of a natural clock that could be used to date the age of rocks and minerals. These radioactive clocks start ticking at the time that molten rocks solidify or crystallize. One concept that scientists developed in order to use these radioactive clocks was the concept of half-life. Though it is impossible to tell when an individual radioactive atom will disintegrate, knowing the constant rate of decay, it is possible to say that after a certain amount of time, half of the atoms in a radioactive sample will have transformed themselves into their daughter element. And after another equal amount of time, half of the remaining radioactive atoms will decay, and so on. As each half-life passes, a sample of radioactive material decays in smaller and smaller proportions, but that proportion is always half of what remains of

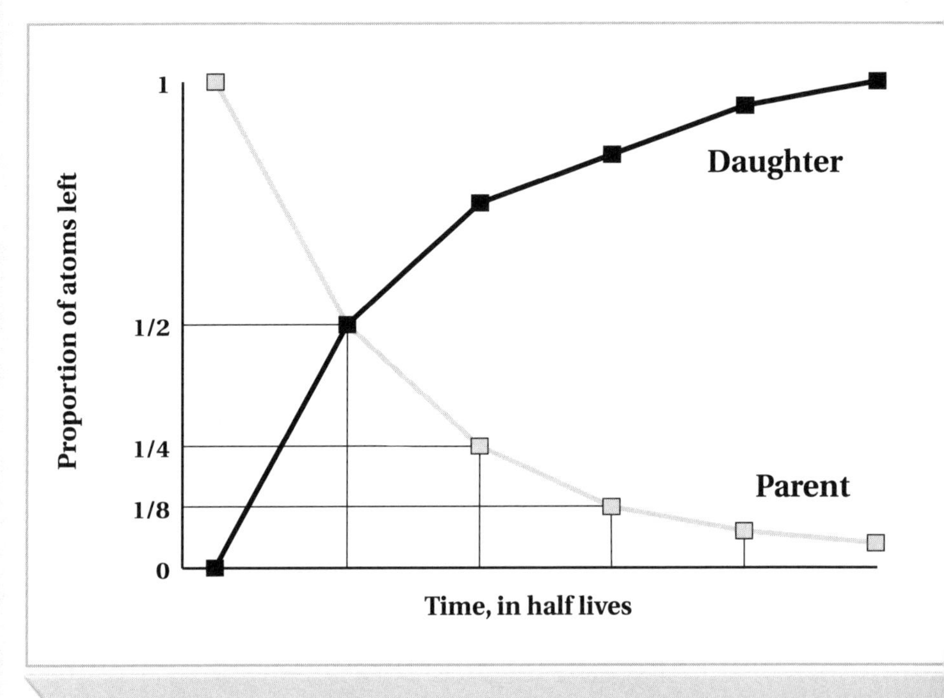

If you know the constant rate of decay, it is possible to determine that, after a certain amount of time, half of the atoms in a radioactive sample will transform into their daughter element.

the radioactive element at any point. For example, the half-life of uranium 235 is over 700 million years. After 700 million years have passed, half of the atoms in a sample of pure uranium will be atoms of lead. After another 700 million years have passed, half of the remaining uranium atoms will turn to lead, and the whole sample will be 75 percent lead.

Using the uranium-to-lead decay rate, Bertram B. Boltwood was able to calculate the ages of more than forty mineral samples, and came up with rocks aged from 400 million to 2.2 billion years. In the 1920s, American astronomer Henry Norris Russell (1877–1957), better known for his work in stellar evolution, also measured the ratios of uranium to lead in mineral samples and came up with an age for the earth's crust of about 4 billion years. Another early pioneer who used uranium-to-lead methods was British geologist Arthur Holmes (1890–1965). In 1927 he calculated a maximum age for the earth of 3 billion years. Two years later, Rutherford would try his own calculations using the decay of uranium to lead in the earth's crust, and arrive at a result of 3.4 billion years.

Estimates for the age of the earth were increasing. Geologists and evolutionary biologists now felt comfortable that the earth had been around for a long enough time to allow common geological processes to shape it and for natural selection to create the complex organisms that lived on it. Baron Kelvin had died in 1907, and never lived to see how the phenomenon of radioactivity would invalidate his own attempts at absolute dating using the laws of thermodynamics. If he had lived to see such advances, he probably would have applauded, for the growth of science was more important to him than his own accomplishments, which were many.

As science and technology improved, radiometric dating techniques would become ever more reliable, and many researchers would test new methods over and over again and compare the results with their first findings. Arthur Holmes would recalculate the age of the earth again in 1946 and 1947, twenty years after his first calculations. By the time of Holmes' death, most geologists had accepted a figure for the age of the earth of 4.6 billion years, and that figure has held up right through the present day. So sophisticated have radiometric dating techniques become that the potential error associated with this 4.6 billion year figure is only one million years either way.

RADIOACTIVE ISOTOPES

There were, however, some limitations to radiometric dating if the decay of heavy elements like uranium were used as age markers. Uranium and similar elements decay very slowly. With a half-life of more than 700,000 years, uranium might be very useful in determining the age of rocks formed billions of years ago. It was not, however, a fine enough measure for events in the relatively recent past, that period of time during which modern life forms evolved and humans appeared on the earth. This problem was solved by the work of Rutherford's assistant Frederick Soddy and American chemist Willard Frank Libby (1908–1980).

In the course of his work with Rutherford on radioactivity, Soddy had detected many elements with similar chemical properties but different atomic weights. He named these variants isotopes. The discovery of the neutron by James Chadwick (1891–1974) in 1932 led to the deduction that the various isotopes of a

Parent Isotope	Daughter Isotope	Half-Life (Years)
Carbon-14	Nitrogen-14	5,730
Uranium-235	Lead-207	710,000,000
Potassium-40	Argon-40	1,300,000,000
Uranium-238	Lead-206	4,500,000,000
Thorium-232	Lead-208	15,000,000,000
Rubidium-87	Strontium-87	47,000,000,000

Elements with similar chemical properties but different atomic weights are isotopes. Radioactive isotopes decay at known rates

particular element all had the same number of protons in their nucleus, but different numbers of neutrons. Some of these isotopes were also radioactive.

Isotopes are fairly rare. In an ordinary sample of water, for example, most of the water molecules contain hydrogen atoms with only one proton in the nucleus. But a very small number of these water molecules contain a hydrogen atom with a proton and a neutron in the nucleus, and an even smaller number contain a hydrogen atom with a proton and two neutrons. All these molecules exhibit the same chemical behavior, but the isotopes of ordinary hydrogen, known respectively as deuterium and tritium, are heavier than ordinary hydrogen atoms. In a sample of unrefined uranium ore, most of the atoms are U238 atoms, and only a few are the more radioactive U235 atoms, with three fewer neutrons in the nucleus.

During World War II, Willard Frank Libby had worked on the highly secret Manhattan Project to develop an atomic bomb. His specific work dealt with the problems of separating uranium isotopes, because pure U235 was needed to make the bomb work. After the war, he became interested in the properties of another isotope, carbon-14, that had been isolated in 1940. Carbon-14 had two more neutrons in its nucleus than the more common carbon-12. Carbon-14 was radioactive, decaying at a constant rate like other

radioactive elements and it had a half-life of only a little more than 5,000 years.

Radioactive carbon-14 is continually being formed in the atmosphere as it is bombarded by cosmic rays. Therefore, a small number of the carbon dioxide molecules in the atmosphere contain radioactive carbon-14 atoms. Libby understood that atmospheric carbon dioxide was continually being absorbed by living plants, and was also present in the tissues of animals that ate those plants. The amount of radioactive carbon-14 in plants and animals was kept at a constant level as long as the plants and animals were alive to absorb or release CO_2 molecules. But when these creatures died, Libby realized, the intake of new radioactive carbon atoms ceased, and the ones already absorbed were no longer released but began to decay at a steady rate. Here was a new type of radioactive clock that because of its short half-life was accurate for relatively small periods of time. And it could be used to measure the age of all organic

he amount of radioactive carbon-14 in living things is constant while they re alive, but the intake stops when they die, and the atoms already bsorbed begin to decay. By measuring how many are in an object, ts age can be determined. Carbon dating has measured objects uch as the Dead Sea Scrolls, above.

materials, like pieces of wood or paper or mummies or the bones of animals that had died relatively recently. Libby had the technique of carbon-14 dating fairly well worked out by 1947. It has been used to date objects no more than about 70,000 years old. The technique has helped to date the Dead Sea Scrolls and the settlement of the Americas by Native Americans.

THE MASS SPECTROMETER

Radiometric dating was further advanced with the invention of the mass spectrometer in the 1930s. A regular spectrograph can identify the composition of materials by heating them to incandescence and analyzing their light emissions. This reveals information about the electron structure of the atoms and thereby identifies elements by their chemical properties. But the nucleus of the atom remains unaffected by heat. No information can be obtained with this method that reveals the structure of the atomic nucleus. All isotopes, therefore, behave exactly alike when analyzed in this way and cannot be told apart.

The mass spectrometer works differently. First a substance is ionized, that is, electrons are removed from its atoms so that each atom is no longer electrically

The mass spectrometer can detect differences in mass as small as a single neutron. It is used to identify the various isotopes of fundamental elements.

neutral but possesses a positive charge. Then these atoms are propelled past electromagnetic fields. The ionized atoms are influenced by the electromagnetic field and their path through the detector is curved. The degree of curvature depends upon the atom's mass, which is a function of how many protons and neutrons it contains. The heavier the atom, the harder

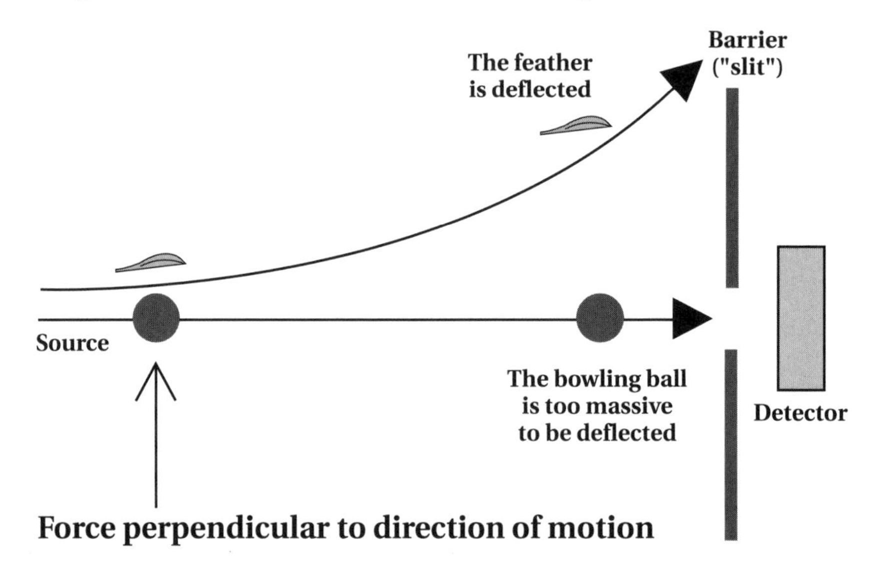

Ionized atoms are propelled past electromagnetic fields of a mass spectrometer. The electromagnetic field causes the atoms' path to curve through the detector. The amount of curvature is dependent upon the atom's mass, which depends on the number of protons and neutrons it contains. The heavier the atom, the harder it is to bend its path through the device.

it is to bend its path through the device. The mass spectrometer can detect differences in mass as small as a single neutron, and therefore the mass spectrometer can identify different isotopes of the fundamental elements. The mass spectrometer was used to identify more than 200 new radioactive isotopes with a variety of different decay rates, and these have proved very useful in determining the age of different materials. Potassium-40, for example, has a half-life of 1.25 billion years, and since potassium is an element taken up by living things, this isotope can be used to identify some fairly old fossils.

Geochronology

Today the tools available to scientists to measure the age of the earth are accurate and reliable enough to give this whole field of study its own name, geochronology. Radiometric dating is still the most widely used and trusted method for determining the age of rocks and fossils, though it does produce some inaccuracies.

In radiometric dating, scientists must assume that the quantities of a radioactive isotope and its daughter element have remained fixed in a rock sample and that the total number of their atoms has not declined or increased over the ages. Given that the time frames involved are often hundreds of thousands to billions of years, this is highly unlikely. It is certainly not true when potassium-40 decays into argon-40. Argon is a gas and gas molecules can escape from rocks. This will change the ratio of parent to daughter elements and result in errors of measurement. Earth, too, is geologically active, with new crust forming from the planet's molten interior and replacing sections of old crust. Each time that rocks melt and reform, their radioactive clocks are reset, so that some of the oldest rocks no longer exist. Furthermore, the contamination of the earth's atmosphere with the radioactive by-products of above-ground nuclear explosions during the 1960s and 1970s may have introduced further inaccuracies.

Nevertheless, radiometric dating has produced some fairly reliable age estimates for rock strata. To date, on every continent, rocks have been found that prove that our planet is indeed billions of years old. Many rocks dating back 3.4 to 3.6 billion years have been found in such regions as the Great Lakes area of North America, in the northern part of Western Australia, in Venezuela in South America, and in Swaziland in southeast Africa. Rocks that may be 3.7 billion years old have been found in Minnesota and Michigan. In western Greenland, there are some rocks that may be 3.8 billion years old. In the Northwest Territories, a region of Canada, near Great Slave Lake, the deepest body of water in North America, geologists have

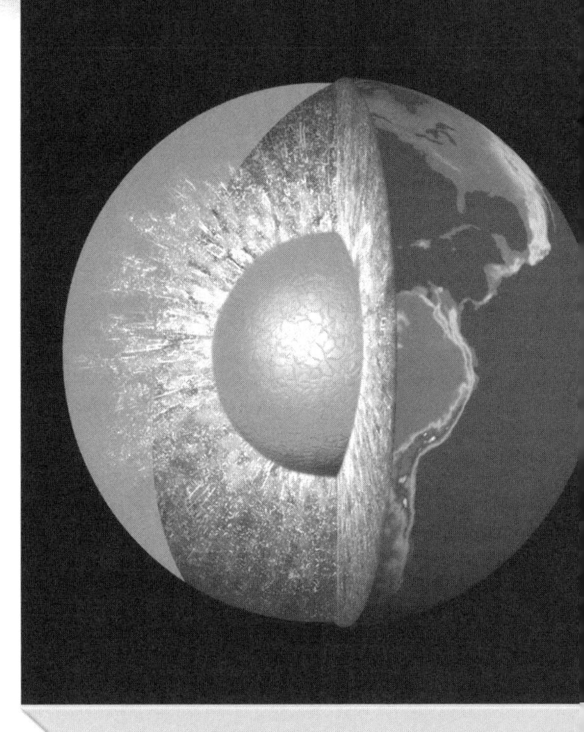

Molten rock boils beneath the surface of the earth and constantly refreshes its crust, making geochronology difficult

The contamination of Earth's atmosphere by radioactive by-products of above-ground nuclear testing has made it difficult to accurately determine the real age of rock strata.

found ancient metamorphic rocks called gneiss that may date back about 3.9 billion years. Rocks of similar age have also been found in Antarctica at Enderby Land.

According to the U. S. Geological Survey, the oldest minerals on Earth are zircon crystals found in sedimentary rocks in Western Australia. Radiometric dating has indicated that these crystals may be as much as 4.2 billion years old. As sedimentary rocks are formed by a consolidation of materials from other sources, these crystals must have eroded from even older rocks. To date, the original sources of these crystals have not been located, but once they are found, we may find them to be more than 4 billion years old.

EXTRATERRESTRIAL EVIDENCE

Today the determination of our planet's age is no longer entirely dependent upon the measurement of the age of rocks on the earth. Rocks have been brought back from the Moon, and geologists can also use radiometric dating to determine the age of meteorites. The current theory of the origin of the solar system, which was actually first proposed by both German philosopher Immanuel Kant (1724–1804) and French astronomer Pierre Simon Laplace (1749–1827), suggests that it formed from a huge, rotating, contracting cloud of hydrogen and helium gas laced with dust particles. According to this theory, all of the planets came into being at about the same time. In fact, everything in the solar system is believed to have come into existence at roughly the same time, including the Moon and the planetary bodies that produce meteorites.

This means that if geologists can find other materials in the solar system that are even older than the rocks on Earth, it is valid to use this evidence to infer the age of the earth. The earth, remember, is a geologically active planet that continually destroys and recycles its oldest rocks. The Moon, however, is now known to be geologically inactive, preserving its oldest structures unless they are altered by bombardment by meteorites, asteroids, and other debris from space. The origins of the Moon are not generally agreed upon. It could have been an independent world eventually caught by Earth's gravity. It could have been part of another planet or it could have been born from the collisions of smaller satellites that once circled Earth. In any event, it formed very early in the history of the solar system.

Moon rocks were collected during six NASA Apollo missions from 1969 to 1972. The Moon appears to be composed mainly of basalts, breccias, and regolith material. Basalts are igneous rocks formed from the solidification of molten lava, composed of minerals like pyroxene and plagiolase. Basalts are dark gray in color. When you look up at the Moon, the dark areas are composed of basalt. Breccias are fragments of other rocks fused together by the impact of meteorites. Regolith is the loose, soil-like material on the lunar surface, also formed from the collisions of meteorites and asteroids, as this material is thrown out of impact craters. Radiometric dating has indicated ages of 4.4 to 4.5 billion years for the oldest Moon rocks, with very few of them less than 3.5 billion years in age.

This 4.4 to 4.5 billion year age range is confirmed by the radiometric dating of meteorites. Meteorites are believed to be fragments of material from asteroids and

Moon rocks were collected during six NASA Apollo missions from 1969 to 1972. Radiometric dating indicated that the oldest rocks are 4.4 to 4.5 billion years old.

comets that have fallen to Earth from space. Their composition varies from stony to metallic. They appear to be among the oldest objects in the solar system.

One such iron meteorite, located in Canyon Diablo thirty-five miles east of Flagstaff, Arizona, left an impact crater 600 feet deep and almost one-quarter of a mile in diameter. The impact was so forceful, generating

such intense heat and pressure, that it may have been the cause for the diamond content in the meteorite. Numerous radioactive clocks were used to date the Diablo Canyon meteorite, including a method that compared its lead isotopes with similar isotopes in samples of terrestrial rocks from various sites in Canada, Australia, and South Africa. Calculations showed that the earth has probably been around as a solid body for the last 4.5 billion years.

It is hard to imagine such a great span of time, or to imagine that science can even be sure that this is the true age of the earth. Based on measurements of the age of terrestrial rocks, however, this age appears defensible, and it is in agreement with measurements of the ages of lunar rocks and meteorites, both 4.6

Meteorites are thought to be bits of material from asteroids and comets that fall to Earth. Their composition varies from stony to metallic, and scientists think they are among the oldest objects in the solar system.

Astronomer Andrew Douglass introduced dendrochronology, the measurement of the growth of tree rings, to determine the age of the wood. Each year, a tree grows a new ring, and counting the rings gives an indication of how old the tree is.

billion years old. The search to confirm or revise this age continues. New ideas and new methods will continue to drive science. Some day, with the help of other sciences such as physics and astronomy, geologists may be able to estimate the age of the earth to within a few thousand years. Or they may find that the current age is billions of years too short.

Other methods of measuring the age of materials have been developed, though they have severe limitations and are usually only used in special situations or when radiometric dating can confirm their results. We've already mentioned efforts to measure rates of sedimentation. Another technique is based on the discovery that the chemical hydroxypatite, found in bones, gradually changes into the chemical fluorapatite when those bones are buried. This technique has been used to date human fossil remains. Geologists also study the alignment of magnetic minerals in rocks, because it is now known that in the distant past the earth's magnetic field has flip-flopped and the North and South Poles have frequently reversed positions.

In the 1920s, American astronomer Andrew Douglass (1867–1962) introduced the technique of dendrochronology, the measurement of the growth of tree rings, to determine the age of both living and dead specimens of wood. Each year trees add a new growth

ring to their trunks, and the size of the ring also indicates whether the climate was favorable for plant growth. Obviously the method is limited to measuring only the very recent past, and the technique has been used as a check on some of the dates produced by carbon-14 dating methods.

The growth of marine coral has also been used to study the earth's age in recent years. Corals are tiny animals that thrive in colonies attached to rocks and other objects in shallow, tropical seas. They secrete a skeleton of calcium carbonate, which builds up a tiny new layer or ridge each day. Many coral reefs today are under severe environmental stress and are not suitable for making these measurements, but there are still a few coral beds in the seas that can be used for

Marine coral can give clues to the earth's age. Coral produces a calcium carbonate skeleton, which builds up a new layer or ridge each day. Many coral reefs today are under severe environmental stress, however.

this purpose. Studies have found that in recent times corals have been laying down about 360 fine skeletal ridges a year, as one would expect, but some older corals show between 385 and 410 ridges. This is further confirmation of the belief that the earth's rotation has been slowing down as a result of gravitational interaction with the Moon, and that in the prehistoric past Earth was spinning more rapidly and there were more days in each year.

CONTINENTAL DRIFT

We're going to take a look at what is known about the ancient history of the earth, but first a word of caution. It has been said several times that the geological record must remain incomplete because the earth is a geologically active planet, but what does that mean? To explain this, we must go back to an idea first proposed by German geologist Alfred Wegener (1880–1930) in 1912. Wegener studied the shape and position of all the

Alfred Wegener developed the theory of shifting continents.

continents on the globe and noted that, if he could move them, they would almost fit together like the pieces of a jigsaw puzzle. The concave west coast of Africa seems to fit snugly into the convex east coast of South America. The west coast of the Americas seems to fit into the curve formed by the east coast of Asia running from Japan south to Australia. Separated by the Mediterranean, the north coast of Africa seems to fit snugly into the southern coast of Europe.

Wegener suggested that in ancient times all the land masses were part of a single super continent, which he called Pangaea, which means "all lands."

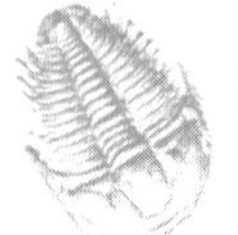

Wegener believed that over immense periods of geologic time the continents had split apart and moved to their present locations, a theory known as continental drift. He also suggested that continents do not always drift apart, but sometimes come together, pushing into each other with enormous force—enough force to crumple the crust of the earth and create mountain ranges. Unfortunately, Wegener could suggest no geological mechanism to move such huge land masses, and his theories were dismissed.

In 1929, however, Wegener's ideas were taken up by English geologist Arthur Holmes, who had earlier proposed that radioactive heat generated within the earth invalidated Baron Kelvin's notion of a young planet. Holmes now suggested that the radioactive heat within the earth was sufficient to melt rocks or at least allow them to flow like soft plastic, and that the interior of the earth underwent thermal convection. At many points around the earth, heated material flowed toward the surface, cooled and solidified into hard crust, and then sank back into the depths to be reheated. A circular current of molten rock was created that could carry the attached continent on top of it and move it horizontally across the surface of the earth.

Much research was done on these ideas beginning in the 1960s, and continental drift is now considered a valid scientific concept, though it is now called the science of plate tectonics. The surface crust of the earth is made up of a series of broken plates, like a cracked egg shell, "floating" on top of a hot, plastic layer of rock. New molten rock pushes its way to the surface and forces the crust to move laterally. This usually occurs along fault lines under the sea known as mid-oceanic ridges, great undersea mountain ranges formed by the extrusion of molten rock. From these ridges, scientists have measured the outward spread of the sea floor. The movement, of course, is measured only in inches per thousands of years. When the spreading sea floor reaches the continental shelves, it is said to be subducted, that is, it sinks back down under the

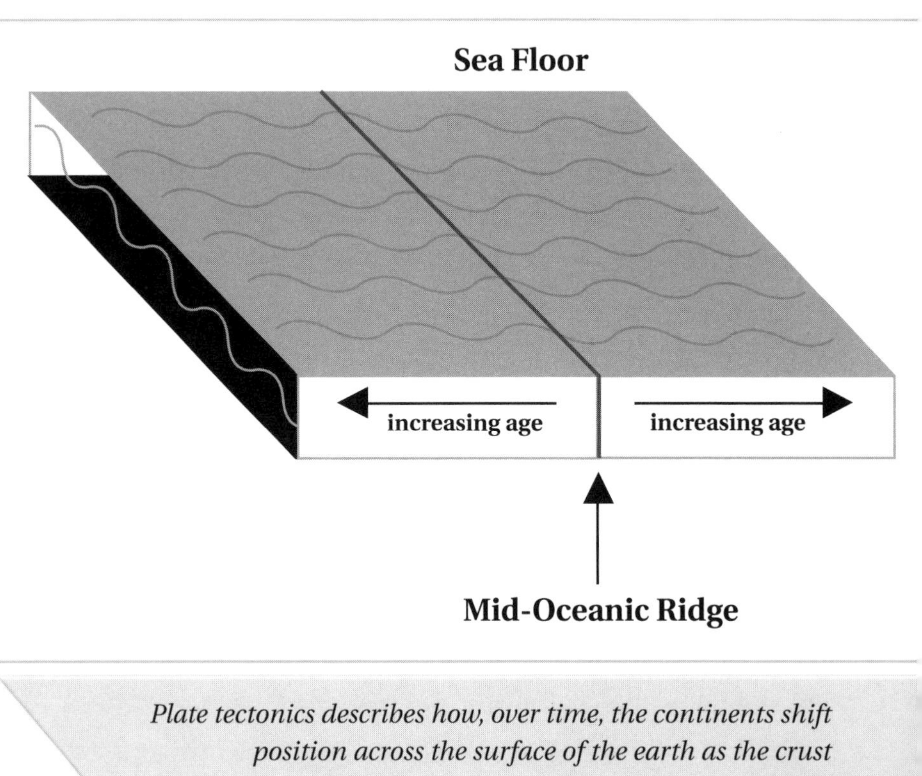

Plate tectonics describes how, over time, the continents shift position across the surface of the earth as the crust of the earth renews itself.

continents into deep sea trenches, only to repeat the cycle over milions of years.

In this way, the continents shift position across the surface of the earth over the ages and the crust of the earth renews itself periodically. Wegener's ideas were vindicated, and we now have a new view of Earth as a planet that actively transforms itself over time. The significance of this for our purposes is that old rocks disappear. They return to the interior of the earth and are melted down and replaced by new, younger rocks. Much of the record of the history of the ancient earth is gone. We know a great deal, but many details of the early natural history of the planet are lost to us forever.

WHAT WE KNOW

We are fairly certain that the earth formed about 4.6 billion years ago. As mentioned earlier, there are a few locations where rocks this old have managed to survive, and the dates are confirmed by the ages of other objects in the solar system like meteors and asteroids, which must have formed about the same time. But then comes a vast period of uncertainty known as the Precambrian era, which stretches from 4.6 billion years ago to about 540 million years ago. This era comprises nearly 80 percent of the earth's history, from the birth of the planet to the first appearance of

complex, multicellular forms of life. Very few fossils have been preserved in Precambrian rock because it was formed under volcanic conditions. The earliest fossils were soft-bodied creatures anyway, and only hard bones and shells are preserved in the fossil record. Nevertheless, evidence of the first unicellular forms of life is preserved in structures called stromatolites, small mounds of minerals that were apparently secreted by living cells. These appear to date the origin of life at about 3.5 billion years in the past. Single-celled creatures multiplied and evolved rapidly and spread throughout the oceans. They were plant cells, and they drew carbon dioxide out of the atmosphere and added oxygen to it.

About 540 million years ago an event known as the Precambrian explosion occurred, when many new forms of life appeared, taking advantage of the enriched oxygen atmosphere, and living creatures moved onto land. This was the beginning of the Paleozoic era, the age of ancient life, which lasted from

Stromatolites resulted from primitive life forms that first existed on Earth 3.5 billion years ago. The dome-shaped structures are formed by single-celled organisms called cyanobacteria.

544 million years ago to 245 million years ago. During this time fish appeared in the oceans and insects, amphibians, and reptiles appeared on the land. Coal was formed deep in the earth from the decayed remains of plants. At the end of the Paleozoic era, some mysterious event occurred that wiped out 90 percent of the various species living in the sea.

The Mesozoic era, the middle era of life, followed. It lasted from 245 million years ago to 65 million years ago, and includes the Triassic, Jurassic, and Cretaceous periods. Dinosaurs evolved during the Triassic period and thrived during the Jurassic period. The Jurassic period, incidentally, is named for the Jura Mountains between France and Switzerland, where rocks from this time were first studied. At the end of the Cretaceous period there was another mass extinction of species that wiped out the dinosaurs and created ecological niches where mammals could thrive.

CATASTROPHISM RETURNS

There is an interesting theory regarding the cause of the Cretaceous mass extinction. In the 1980s, American physicists Luis and Walter Alvarez discovered a rather high concentration of the element iridium in a layer of sedimentary rock that formed at the end of the Cretaceous period. Iridium is rare in

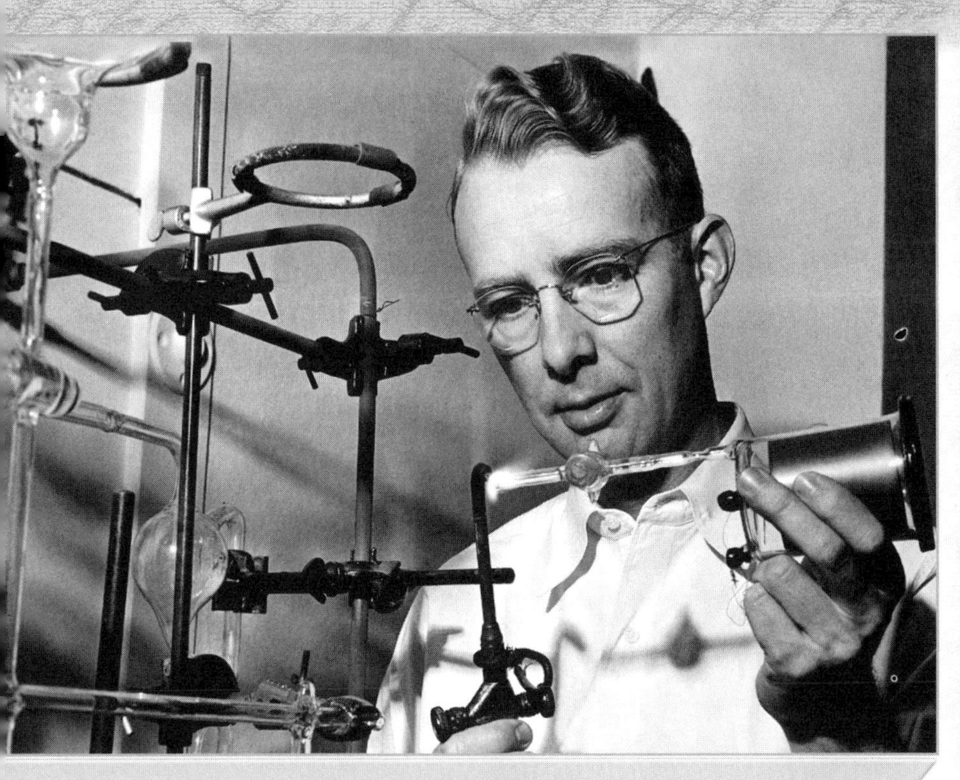

Luis Alvarez (seen here) *and his son Walter proposed that a massive asteroid hit the earth, shooting up a huge cloud of dust and debris that blocked sunlight for many years and cooled the earth's climate, killing off many species.*

the earth's crust, but it is found in greater concentrations in asteroids and extraterrestrial rocks. The father and son team proposed that the extinction of the dinosaurs was caused by the impact on the surface of the earth of a massive asteroid. The asteroid's impact is believed to have sent up a huge cloud of dust and debris that blocked a portion of sunlight for

John R. Foster
© 2/95

many years and cooled the earth's climate, killing off many creatures.

This theory is now accepted by many geologists, and in fact there is some evidence to prove that the asteroid impact occurred near the Yucatan peninsula of Mexico. The theory has many interesting implications. It means that the natural history of the earth is sometimes determined by accidental and extraterrestrial events that cannot be predicted by study of the earth's dynamic geologic processes. Today, in fact, geologists believe in a modified version of Lyell's uniformitarianism. Although most geological and evolutionary change occurs gradually, and uniformitarianism is the correct answer to the catastrophism proposed by theologians, there are periods when rapid catastrophic changes do occur and bring about major events such as mass extinctions.

Scientists theorize that the extinction of the dinosaurs created enough of an ecological opportunity for smaller mammals to thrive and eventually evolve into human beings.

There is also a lively debate now taking place among biologists about how rapidly evolutionary changes take place in living species. American paleontologists Stephen Jay Gould (1941–) and Niels Eldridge (1943–) have proposed a modification of Darwin's theory called punctuated equilibrium that calls for more rapid periods of change, but the theory is controversial. Gould and Eldridge believe that there are long periods of time when species are stable and unchanging, interspersed with shorter periods of dramatic environmental change when rapid adaptation becomes necessary in order for creatures to survive. Critics like British paleontologist Richard Dawkins (1941–) would concede that rapid evolutionary change can take place, but emphasize that the basic mechanism of natural selection, small changes over very long periods of time, can account for the appearance of complex

This crater was formed by a meteor impact millions of years ago.

Carl Sagan theorized that a nuclear war would cause a nuclear winter that would devastate the earth's climate. He also popularized science through his PBS television series Cosmos.

organisms like ourselves without the need for a more elaborate theory. In any case, neither Gould nor Eldridge nor Dawkins would question that the fossil record is solid evidence for a very ancient Earth.

The asteroid collision theory was also used during the 1980s to demonstrate what would happen to the planet in the aftermath of a nuclear war.

Astronomer Carl Sagan (1934–1996) and Gould argued that nuclear explosions might lift enough dust and debris into the atmosphere to block the Sun's light and alter the earth's climate. The effect on agriculture would be the most devastating, since it would become too cold and dark to grow crops, and without agriculture human civilization would collapse. This was known as the nuclear winter scenario.

In any event, after the Mesozoic era comes the era we live in, the Cenozoic era, the age of recent life, which began 65 million years ago. The Cenozoic era is sometimes called the era of mammals. The extinction of the dinosaurs created the ecological space and the freedom from large predators that enabled small mammals to grow larger and dominate the landscape. Human beings evolved somewhere between four and two million years ago, and we have already survived several reincarnations of our species: *Australopithecus*, Neanderthal, and *Homo sapiens*. We humans live in the Holocene epoch of the Quaternary period of the

Cenozoic era. What awaits us in the future, assuming that we don't cause our own destruction through war or mismanagement of our environment?

That is difficult to answer, but our planet is still meteorologically and geologically active, and so human civilization faces the likely prospect of a changed Earth in the distant future. Another ice age is a distinct possibility. Continents will continue to change shape and shift their locations. Given enough future time, the likelihood of a catastrophic collision with a body from space increases. In another 10 billion years or so, our Sun will run out of nuclear fuel and undergo

Our species, Homo sapiens, *has evolved over two to four million years.*

changes that will probably end all life on Earth. And yet these dramatic changes are overshadowed in our minds by the more immediate problems that we have brought upon ourselves. Will we have enough clean water to drink and clean air to breathe a mere hundred years from now? Will our atmosphere continue to protect us from dangerous ultraviolet radiation? Will there be enough food for the projected increase in world population? How will we run our civilization when there is no more oil?

The history of science is the history of changing perspectives. Humans no longer see themselves as favored creatures at the center of the universe. Nor are we favored creatures in the long history of our planet. Bacteria and insects have much better survival mechanisms than humans, in spite of our intelligence. Will we be around to witness the great changes that will occur to our planet in the distant future? If we can survive the next century or two, we may earn the right to ask that question.

Glossary

asteroid A small planetoid that orbits the Sun. Some meteorites were once part of asteroids. Most asteroids are found in a ring or belt between the planets Mars and Jupiter, and are believed to be the remains of a planet that never formed properly. Gravitational forces sometimes push an asteroid out of its belt, and it can become a threat to other planets.

atmosphere The gaseous envelope that surrounds a planet.

crater A circular depression on the surface of a planet caused by the impact of an asteroid or meteor.

dendrochronology The measurement of the growth of tree rings to determine the age of specimens of wood.

eon The longest duration of geological time.

era A division of time included within the span of an eon.

fossil The remains of an ancient organism found in sediment or rock. Only hard tissues such as bones and shells form fossils.

fossil succession The principle that in a vertical cross section of rock strata the older fossils are at the bottom and the younger fossils are at the top.

half-life The time it takes for half of the radioactive atoms in a sample of material to decay into their stable daughter element.

igneous rock Rock formed by the cooling of molten material.

isotope A variant of a chemical element that has the same number of protons in the nucleus of its atoms, and therefore the same chemical properties, but a different numbers of neutrons.

lava Molten rock on the surface of the earth.

mass spectrometer An instrument that can identify elements and their isotopes by making precise measurements of the mass of their atoms.

metamorphic rock Rock formed by exposure to extremely high temperatures and pressures.

natural selection Charles Darwin's theory that the environment puts pressure on species to adapt, and those who are the most well-adapted, or "fittest," survive.

period In geology, a division of time included within the span of an era.

plate tectonics The modern version of the theory of continental drift, in which the earth's land masses are in motion over hot molten layers just below the earth's surface.

radioactive decay The spontaneous disintegration of certain heavy atoms into one or more atoms of different elements.

radiometric dating The determination of the age of rocks by measuring the progress of the decay of the radioactive elements within them.

strata Layers of rock.

tides The movement of the oceans in response to the gravitational force of the Moon.

For More Information

The Berkeley Geochronology Center
2455 Ridge Road
Berkeley, CA 94709
(510) 644-9200
Web site: http://www.bgc.org

USGS Information Services
United States Geological Survey
Box 25286, Building 810
Denver Federal Center
Denver, CO 80225
(303) 202-4700
Web site: http://pubs.usgs.gov/gip/geotime

WEB SITES:

Geological Society of America
htpp://www.geosociety.org

Kids Web: Geology
http://www.kidsvista.com/sciences/geology.htm

University of California, Berkeley, Geology Wing
http://www.ucmp.berkeley.edu/exhibit/geology.html

Volcano World
http://volcano.und.nodak.edu

For Further Reading

Busch, Richard M., Edward J. Tarbuck, and Frederick K. Lutgens. *Earth: An Introduction to Physical Geology: Study Guide.* Upper Saddle River, NJ: Prentice Hall, 1999.

Cattermole, Peter. *Building Planet Earth: Five Billion Years of Earth History.* New York: Cambridge University Press, 2000.

Dalrymple, G. Brent. *The Age of the Earth.* Stanford, CA: Stanford University Press, 1991.

Garcia, Frank A., and Donald S. Miller. *Discovering Fossils: How to Find and Identify Remains of the Prehistoric Past.* Mechanicsburg, PA: Stackpole Books, 1998.

Haber, Francis C. *The Age of the World: Moses to Darwin.* Westport, CT: Greenwood Press, 1978.

Lewis, Cherry. *The Dating Game: Searching for the Age of the Earth.* New York: Cambridge University Press, 2000.

Norton, O. Richard. *Rocks from Space: Meteorites and Meteorite Hunters.* Missoula, MT: Mountain Press, 1998.

Prothero, Donald R. *Interpreting the Stratigraphic Record.* New York: W. H. Freeman, 1990.

Stanley, S. M. *Earth and Life Through Time.* 2nd ed. New York: W. H. Freeman, 1989.

Index

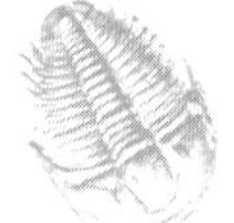

Credits

ABOUT THE AUTHOR

Charles J. Caes is the author of more than a dozen books on science and other subjects. His previous work for Rosen was *The Young Zillionaire's Guide to the Stock Market.* He lives in Virginia with his wife Karen.

PHOTO CREDITS

Cover © SuperStock; cover insert © James L. Amos/Corbis; p. 8 © Omni Photo Communication, Inc./Index Stock; pp. 13, 36, 79, 87 © Bettmann/Corbis; p. 17 © David Parker/Photo Researchers; pp. 19, 22, 23, 24, 27 © Northwind Picture Archives; p. 20 © Beverly Factor/Index Stock; p. 28 © Archive Photos; p. 30 © Academy of Natural Sciences of Philadelphia/Corbis;

pp. 42, 59, 61, 74, 76 © SuperStock; pp. 43, 66, 67 © PhotoDisc; p. 71 © Corbis; p. 72 © Jack Zehrt/FPG; p. 85 © Jonathan Blair/Corbis; p. 88 © John Foster/Photo Researchers; p. 90 © Hulton/Archive; p. 92 © Santi Visalli Inc./Archive Photos; pp. 94, 95 © BSIP Agency/Index Stock; diagrams on pp. 52, 62, 82 by Geri Giordano.

DESIGN AND LAYOUT

Evelyn Horovicz